THE NATIONAL
BIRDS OF PREY

Published by The National Birds of Prey Centre
Newent, Gloucestershire
Tel: 0870 990 1992
Email: jpj@nbpc.demon.co.uk
www.nbpc.co.uk.

© The National Birds of Prey Centre

Written by Jemima Parry-Jones MBE

Cartoons by John Crooks

Photographs by lots of friends!

Cover illustration by Simon Rumble

Designed and Printed by Stoate & Bishop (Printers)

ISBN 0 9534122 0 2

THE NATIONAL BIRDS OF PREY CENTRE
www.nbpc.co.uk
CONSERVATION THROUGH EDUCATION, CAPTIVE BREEDING, RESEARCH, REHABILITATION

WALK, OBSERVE AND ENJOY 2

THIS GUIDE BOOK 6

WHAT DO THE BIRDS DO HERE 6

WHAT EXACTLY ARE BIRDS OF PREY 6

WHAT IS THE NATIONAL BIRDS OF PREY CENTRE 7
    The Flying Birds 9
    The Breeding Birds 9
    The Flying Demonstrations 10

WHO IS WHO? 10

DIURNAL BIRDS OF PREY 11
    The New World Vultures 11
    The Osprey 11
    The Secretary Bird 12
    The Hawks and Eagles 12
    The Falcons and Caracaras 16

NOCTURNAL BIRDS OF PREY (OWLS) 19
    The Barn Owls 19
    The Rest of the Owls - sometimes known as the Eared-Owls 20

MIGRATION 22

IDENTIFICATION 23

SOME OF THE OTHER WORK AT NBPC 24
    Work Experience and Training 24
    Rehabilitation 24
    Assistance abroad 24
    Advice 24

THE BREEDING SUCCESSES 25
    The Breeding Barns 25
    The resulting young 26

SOME STORIES ABOUT SPECIAL BIRDS HERE 27

QUESTIONS FOR YOU TO ANSWER 28

HOW YOU CAN HELP 29

COURSES & MAIL ORDER 30

FURTHER READING 31

USEFUL ADDRESSES 32

THE NATIONAL BIRDS OF PREY TRUST 33

GLOSSARY 34

MAP 35

This Guide Book

We would like this guide book to be useful to anyone purchasing it, long after your visit to the Centre. Consequently it will include general information about birds of prey, rather than describe the individual birds here in the various pens. We feel that a guide book that gives an understanding of the different family groups of birds of prey, where they live, how they hunt and what problems they face in their habitat, will be more interesting and more beneficial long term. The staff are always available and pleased to answer questions should you have any.

What do the birds do here

We utalise the birds for one of two needs. They are either are a part of the flying team, or are here to pair up and breed. This does not mean that those who fly don't breed, or those that are put in for breeding don't fly. If birds in aviaries show no signs of breeding they may well become part of the flying team, and vice versa. If birds that would normally work on the demonstrations show an inclination to breed, they get given the opportunity. There are a very few birds who are too old to do either job and they have a well earned retirement home here.

The Centre has grown from 12 aviaries to 125, all of which have been rebuilt at least once. We are always planning new aviaries with new designs, although not always able to afford those we need and would like. Almost all the aviaries are on view, only five moulting pens and the quarantine quarters are off view, and this is for the sake of the birds. The hospital area is also off view.

What exactly are birds of prey

Many think a bird of prey or raptor, as they are often called, is a bird that eats meat and catches and kills its own food, which to a certain extent is true. However so do ravens, and most of the crow family, also all the tiny insect eaters like wrens, as well as all the fish eating birds such as pelicans, herons, and gulls.

Tawny Eagle in flight
(Aquila rapax)

Savignys Eagle Owl
(Bubo ascalaphus)

The abilities that makes a bird of prey the specialist that it is, are its feet and its eyesight. The other groups of birds mentioned generally use their beaks to catch their food. The raptors have specialised over millions of years; their feet have developed to become more than just perching, walking or swimming aids - they are powerful enough to use for catching their prey; their eyesight has evolved to be quite superb in some cases. Unusually for birds they have binocular vision and the ability to see in detail, that is far greater than humans.

Almost all the birds you see here will use their feet to catch food - but there are exceptions. The vultures are a little different. The Old World vultures evolved from eagles but during their evolution they altered a little and became birds that instead of hunting live quarry, as do the eagles, spend many hours looking for food that requires nothing more than really good eyesight to find - because they are scavengers and specialise in eating food that is already dead. The New World vultures - all those that come from the Americas rather than Eurasia or Africa, evolved along another route, and aren't really true birds of prey - they achieved what is called parallel evolution and they actually are more related to storks than eagles. However, they have grown to look very similar to the Old World vultures because they live in the same way. There are differences which you can see if you look closely at the vultures here at the centre. (See VULTURES)

Most birds of prey have what is called reverse size dimorphism - which means in this case that

Lanner Falcon
(Falco biarmicus)

the female is larger than the male. This is unusual, in many creatures it is the male that is the larger, but not with birds of prey. In just a few species of raptors the male is larger, and in quite a lot the male is only fractionally smaller. Many species the male is a third to half the size of the female.

The only other birds that you are liable to see here - apart from all the wild garden birds are Ravens, Jackdaws and Chickens. None of these are birds of prey – although all live on meat as well as other things and all use their beaks, not their feet with which to catch food.

What is the National Birds of Prey Centre

We are a specialist zoo holding only raptors and open to the public for a number of reasons. It would be wrong to have this significant collection of birds and not share it with others. At the same time we are able to pass on the knowledge that we have to those who visit, both those as normal visitors or as a specialist student. The birds serve as PR officers for their wild relations, by allowing visitors a close up view and a greater understanding. As much as possible we encourage them to breed. This in turn allows us to learn and understand more about them, which then gives us the ability to do more should any species become in need of assistance in the wild in the future.

The National Birds of Prey Centre started life in 1967 as The Falconry Centre, it was founded by Phillip Glasier and his family. Falconry had been a part of the family's life for three generations. Great Uncle Captain Knight was a falconer, writer, lecturer and film maker. Phillip Glasier also wrote books, did film work and eventually worked for the actor James Robertson Justice as his private falconer before founding the Falconry Centre. In 1983 he retired and Mrs. Parry Jones, his daughter, bought the Centre from him to continue the work he started. Mrs. Parry Jones received an MBE for her services to Bird Conservation in 1999. The Centre received an award 'Heroes of Gloucestershire' for its work in the bird conservation field in 2000. It has also received an award for sustained captive breeding of birds of prey from the Federation of Zoos.

Since 2000 a charity called The National Birds of Prey Trust has been established. This will, once it has funds available, take over some of the work that the Centre now does. Some of its remit will be funding the rehabilitation work with the injured birds that come in; providing free

Lanner Falcon
(Falco biarmicus)

Female Snowy Owl
(Nyctea Scandiaca)

educational packs for visiting schools; training for overseas students; support of insitu projects and many other plans in the future.

The Flying Birds

Most of the trained flying birds live tethered in the Hawk Walk - they are tethered only during their work season, which is usually about 8 months. Once they have finished their assignment for that year, they go into an aviary for a rest and to moult. Most of the birds that are tethered are individuals that don't do well if flown from a pen. Some birds get so excited when they see us that they will fly headlong into the wire and injure themselves. By being tethered, they are not able to injure themselves. They are flown free everyday - as you will see if you watch one or more of the flying demonstrations.

Other birds on the flying team, such as the owls, vultures, caracaras and the secretary bird, all work, but instead of being tethered they are kept loose in pens. They are either carried down to the flying ground on the fist, in a box, or just let out of the aviary door.

For anyone wanting to take up falconry - look at the pens just outside the Hawkwalk that house mainly Eagles. These are pens designed for flying birds being what is called free lofted, and are what we recommend that anyone having just one or two birds should build to house them, although you don't have to use materials quite as heavy or expensive as we do.

The Breeding Birds

The breeding birds are housed in pens in the Barns that you will see built around the place. You will notice that almost all the pens are completely roofed over. This gives the birds a great deal more shelter in the increasingly more extreme weather conditions that we are now having. Each breeding pen has a service

American Black Vulture (*Coragyps atratus*)

corridor where we can feed, clean and monitor the birds from in comfort and without disturbing the birds. These pens are designed for the comfort of the birds and ease of management. They do not display the birds in natural habitats, which is something we would like to do next.

At the time of writing we have no 'display aviaries' - pens that are designed specifically to look like natural habitats, rather than designed for breeding and management. This is something that we are planning to do in the future. Two things should be remembered here - most natural habitat exhibits are built for the on-looking visitors, not the inhabitants, our birds are far more interested in watching what goes on around them than whether or not their perch is a real tree, or the bath part of a stream. The other point to remember is that for a serious captive breeding program nest sites, eggs and young have to be monitored and be accessible to keepers. This is not usually possible in 'display or exhibit aviaries'. Therefore when we do get to the stage of building such pens, they will house locally native species or other species that can cope with a less sheltered environment and that we do not mind whether or not they breed.

The pens are cleaned out once a week unless the birds are sitting on eggs or have young babies that would be upset by the disturbance of someone entering the pen for the cleaning process. Once or twice a year we do a major clean - catching up the birds, worming them, checking them over, trimming beaks and talons. While the birds are boxed up we scrub and disinfect the pens from top to bottom, clean the nest ledges and build nests for those would require them. We are experienced nest builders here, we probably build upwards of 40 nests per year. This really does make a difference to our breeding successes. Of course some of the birds ignore the nests, some demolish them completely, some use the materials to build what they consider to be a superior model, and some appreciate our efforts and use the nest as it stands.

We do other things to encourage the birds, we give birds that may be showing signs of breeding wooden eggs to encourage them to lay their own eggs. If they sit on those we will even give them babies and that really makes a difference. It can change a disinterested bird into a dedicated parent in one season.

So if you see a pen looking untidy - just ask one of the staff and you may find that the reason is either eggs or babies in the nest, remembering that our breeding season is very extended, lasting from November to August.

The Flying Demonstrations

There are three flying demonstrations per day. October - March, the winter times, are at 11.30am 1.30pm and 3.30pm. The summer times, April – September are 11.30am 2.00pm and 4.00pm. Each flying demonstration has a completely different team of birds and usually at least half of the birds are different species. As each bird is flown once in the day and then fed, it is a good idea to try and watch at least two demonstrations. Some birds fly better in the summer than the winter and vice versa. For example the Kites who are tremendous to watch are usually flown in the spring and summer months at the 2.00 p.m. demonstration. Whereas the Snowy Owl prefers to fly at the 11.30 a.m. demonstration in the winter. If there are certain birds you want to see - ask one of the staff if and when they are on so you don't miss it, or you can phone to check for further visits.

Who is who?

There are about 250 different species of the diurnal or day flying birds of prey and around 130 species of the nocturnal birds of prey or owls. (It's difficult to give exact numbers of species because not all taxonomists agree). These two taxa or groups are divided into families, and these families are usually put into a certain order, so we are going to follow that order in case you want to use other books to look up more information (See reading list). The families are divided into smaller families and then into individual species.

Birds of prey have evolved to look differently, hunt in different ways at different quarry, and generally take advantage of the differing habitats of the world in which to survive. However, they have similar attributes as well. Superb binocular vision, generally powerful grasping feet, in fact the name raptor means to seize or plunder. They usually live alone or in pairs, although some will live in groups. All feed on meat of some sort, from the tiniest of insects to the largest of mammals, although these are generally eaten as carrion. There are a very few that will feed on vegetable matter as well.

Diurnal birds of prey (Scientific group name **Falconiformes**)

The New World Vultures
(Scientific family name **Cathartidae**)

First of the families are the New World Vultures, you will see several species here at NBPC. This group contains the smallest of all the vultures and the largest - which incidentally is the largest flying bird as well. The American Black Vulture is the smallest, although not the lightest - a little black vulture to be found in many parts of the New World. The King Vulture is the most colorful of all the raptors, and the largest is the Andean Condor, weighing in at up to 28 lbs and a wing span of nearly ten feet. There are seven species of New World Vulture and one of these, the Californian Condor was one of the rarest vultures in the world. This huge bird had dropped down to only 17 individuals left, eventually only in captivity. Thanks to a breeding programme run initially by the San Diego Zoo and the LA Zoo and later joined by The Peregrine Fund, there are now over 150 of these birds, the number increases yearly, young are now being released into safe areas in the wild.

Like all vultures, the New World vultures have specialised in eating carrion (dead animals). Unusually for birds however, some of these American vultures have a sense of smell. With most of them it is only very weak and rudimentary, but the Turkey Vulture has a highly developed sense of smell, by flying slowly over the jungle canopy, this bird can find carcasses hidden by trees and the undergrowth only one hour after the animal has died. Other species of vultures have learned to follow the Turkey Vultures as a way to find food in thick cover. All the vultures are all superb flyers and the smaller ones can soar for many hours on the tiniest of air currents.

King Vulture
(Sarcorhamplius papa)

The Osprey
(Scientific family name **Pandionidae**)

This species, only one in the group has always been placed in its own individual family. It is a very specialist feeder, although there are other raptors and owls that catch and eat fish, none specialises quite as much as the Osprey which hunts and feeds almost solely on live fish. They rarely eat carrion and only very occasionally eat anything other than fish. They are found in every continent around the world except Antarctica, living by inland waters, estuaries and coastal areas. Surprisingly although they are spread so widely, there is little variation, they are a little bigger or smaller depending on where they come from and the colour varies a little, but not significantly considering their range. They have a pronounced angle or crook to their wing shape when they fly which makes them easy to spot. Some of the populations migrate - the ones that come to the UK for example. Others stay on home territories and are more sedentary. They build large nests and lay one to three eggs. When hunting for fish they drop right into the water sometimes disappearing almost completely under water.

The Secretary Bird
(SCIENTIFIC FAMILY NAME **SAGITTARIIDAE**)

Like the Osprey this bird is the only one in its family and it is fairly obvious why - there is no other bird of prey that looks like it or behaves in the same way. Sometimes it is placed before the hawks and eagles and sometimes afterwards. Standing three feet high with legs like a heron this odd bird is found only in Africa below the tropic of cancer. It spends most of its time on the ground walking through the grasslands and bush looking for things to eat. It is famous for killing snakes by stamping them to death, however it actually catches relatively few snakes and the bulk of its quarry is grasshoppers and locusts. It will also eat other large insects, small mammals, birds especially young ones and ground living birds, eggs and the occasional tortoise. The Secretary Bird is a capable flier, with a wingspan of seven feet, and its long legs dangling out behind while it flies. It breeds at the end of the dry season so as to take advantage of all the food around during the wet season.

Secretary Bird (*Sagitarius serpentarius*)

The Hawks and Eagles
(SCIENTIFIC FAMILY NAME **ACCIPITRIDAE**)

This commonly used title is a little misleading as this group covers hawks, eagles, kites, old world vultures, buzzards and many others. It's a huge group with 237 species under the scientific name Accipitridae. This group begins with baza's and kites. The baza's are very odd birds of prey, they are closely related to the kites, tend to be forest birds, live in Africa, South East Asia or Australia. They have fairly weak feet and beaks and they all eat insects, frogs or similar. The kites are probably best known in the UK for the Red Kite, which is the largest of the family. The smallest is the tiny Pearl Kite, which comes from South America and weighs only 80 - 100 grams. Most of the Kites feed on insects, reptiles or small mammals. They all have relatively short legs and like the baza's, weak feet. The largest prey the Red Kite takes is small baby rabbits, but like many of the Kites it mainly eats carrion which leads to it often being thought to kill much larger quarry than it really can. When looking at the kites – notice the foot size - you will see they are small in comparison to the size of the bird - a sure sign of weak feet and small prey items.

In amongst the Kites are the honey buzzards. These are very kite like, but lack the forked tail that marks many of the kites. The honey buzzards have specialised in their prey - they all, as far as we know eat wasp and bee grubs, from the honey comb when they can.

The Hawk Walk

The last of the kites is the Brahminy Kite, this is a fishing kite, which leads nicely to the fishing Eagles. You will see a very good selection of this group at the Centre. Note that many of them have either white tails, or other large areas of white plumage. The largest of all is the Steller's Sea Eagle, this stunning bird is one of the four 'huge' eagles, has a enormous beak and lives on the far north eastern Chinese and Russian coastline. All the fish Eagles, including the best known, the American Bald Eagle and our own White-tailed Sea Eagle, catch fish by snatch lifting them from the surface of the water, but unlike the Osprey they will eat other prey items as well. In fact they will eat almost anything they can find. They are greedy, noisy birds and will happily eat carrion. They are also dramatic flyers, particularly the White-bellied Sea-eagle, which does a dramatic cart-wheeling flight in the wild.

Yellow-billed Kite *(Milvus migrans parasitius)*

Red Kite *(Milvus milvus)*

Next come the Old World Vultures. When you look at them, you will see that unlike the New World vultures, their feet are more eagle-like and you can't see right through their nostrils or nares as you can with the American vultures. The smallest is the Egyptian Vulture, which actually is found in Africa Europe, Asia and India. It is called the Egyptian Vulture because it is seen in the hieroglyphics in the Egyptian Tombs. The largest is probably the European Black Vulture, also known as the Cinerous Vulture. Birds often have more than one name, that is why it is important to know the scientific name as it is unique, so mistakes cannot be made. The Bearded Vulture is a huge bird as well and feeds almost solely on bones. It picks up large ones that it can't swallow and flies up and drops them onto the mountain side to break them into smaller pieces. The vultures in Africa eat more meat than all the other carnivores, such as lions, cheetahs, hyenas and so on, put together. If it were not for the vultures in India and Africa there would be much more disease, as the vultures clear away all the dead and diseased animals.

Female Bald Eagle *(Haliaeetus leucocephalus)*

The Snake Eagles as their name suggests eat snakes among other prey items. They mostly have large heads and eyes and very tough scales on their feet to protect them from bites. All have yellow eyes and none are found in the New World. They have mostly specialised in eating snakes and reptiles, the Bateleur is a bit of an exception as it will take small mammals and eat carrion as well. Most have a short crest of feathers at the back of their head.

There are 13 species of Harriers found around the world. All have a slight facial disc of feathers round the eyes somewhat like, but less marked than that of the owls. All have long broad wings, but a light body weight, giving them the ability to hunt by flying low and slow, again a little like some of the owls, looking and listening for mice and voles and small birds in marshland, coastal, arable farm land and moorland areas. They have long legs and weak feet. In some species the males are a different colour from the females.

Next comes an assortment of raptors, some of which are more closely related to the former sub family - the harriers and some of which are nearer to the accipiters or 'true hawks.' The Gymnogene or African Harrier Hawk, which was first bred in captivity, outside Africa, at the Centre is a large lightweight bird similar to the harrier, but it has a double-jointed ankle, which allows it to bend its foot the 'wrong' way up to 40%. This enables it to take baby birds by putting its leg down hollow trees and holes in cliffs and hooking out the young. Also in this group are the Chanting Goshawks; and the

Verreaux's Eagle (Aquila verreauxi)

Gabar Goshawks, which the Centre bred as a world first. They have a melanistic or black phase, which is very striking and in the wild their nests are often occupied by and coated with the webs of colonial spiders.

The accipiters are the largest genus (related group) of birds not only in birds of prey, but, of all birds - 49 species in all. The Variable Goshawk has the greatest number of subspecies – 23 in the raptors. The accipiters or true hawks, such as the Sparrowhawk and Goshawk, the only two of this group found in the UK, are mainly forest living birds. They vary in size from one of the tiniest of all the diurnal birds of prey - the African Little Sparrowhawk weighing between 74 - 105 grams to the Northern Goshawk, some of which can weigh in at 1509 grams. Short rounded wings, long tails for good steering and for brakes, usually yellow or orange eyes, although not always, and a nervous disposition. The goshawks have thick strong legs and powerful feet, the sparrowhawks have thin, fine legs, long toes and needle sharp talons.

Eurasian Sparrow-hawk (Accipiter nisus)

Moving through another odd grouping which has some wonderful names like the Plumbeous Hawk, the Grasshopper Buzzard, the Semiplumbeous Hawk. Here you will see the Grey Buzzard-eagle and the Harris Hawk. None of these really fit into either the Accipiters or the Buteos (true Buzzards), but instead could be called another link. The Harris Hawk is a very unusual raptor because it is truly sociable. In the most northern part of their range they will live, breed and hunt together in groups and adults will help young. Young will feed other young and birds will work together as a team when hunting.

The buzzards or buteos are another large genus or group. Over the last few years the Common Buzzard which used only to be found in Wales and the west side of Britain has increased dramatically and now can be seen soaring and gliding over much of Britain's farmlands. If you see a large bird about two feet high either sitting on the ground or a post you will most likely have seen one. And that describes many of the buzzards or buteos, they tend to be lazy sedentary birds. Large broad wings and a short tail, they will sit for hours still hunting, or will soar on thermals or on ridge currents. They hunt mainly small mammals, reptiles and feed on a great deal of carrion.

Close to the buteos but usually larger come the eagles. Its interesting to see how it all works and how each group of birds follows on from the last, continuing to change and evolve. There are a number of eagles, some well known, some very rare, all with their powers much exaggerated by mankind. The rarest is probably the Philippine Eagle - a huge

Harris' Hawk (*Parabuteo unicinctus*)

forest eagle that has had much of its habitat destroyed. A similar problem is happening with the Harpy Eagle - another huge forest eagle but this time from South America. The most widely

Northern Goshawk (*Accipiter gentillis*)

Common Buzzard *(Buteo buteo)*

spread of all the eagles is the Golden Eagle found in many parts of the Northern Hemisphere. The largest of the African eagles is the Martial Eagle.

Many of these large birds live in the mountains and open plains of the world. But some have evolved to become more hawk-like and live in the woods and forests of the world. The largest of these is the African Crowned Eagle and the smallest is Wallace's Hawk Eagle. There are eight hawk eagle species that can be found in South East Asia, however, like the Philippine eagle, their days may well be numbered if the primary forests around the world continue to be felled at the rate they are being lost today

The Falcons and Caracaras
(SCIENTIFIC FAMILY NAME **FALCONIDAE**)

The falconidae comprises four different groups - the falcons, the caracaras, the forest falcons and the pygmy falcons.

The caracaras do not look much like the falcons, but there are similarities and the eggs they lay are exactly the same lovely mottled brown colouration. They all come from Central or South America. The Common Caracara is the best documented, several of the caracaras are little understood and have even less information written about them. Two of the species are forest living and like any forest species they are difficult to study. As the forest declines, the habitat for these birds goes and although they are not defined as endangered, it seems logical that if their habitat is endangered then by definition - so are they. Most of the caracaras have long strong legs as they spend a great deal of time on

Philippine Eagle *(Pithecophaga jefferyi)*

Pair of Golden Eagles *(Aquila chrysaetos)*

Tawny Eagle *(Aquila rapax)*

the ground. They are very intelligent birds and although they hunt small mammals and birds occasionally, they are opportunist feeders and will eat carrion and even rotten fruit.

The forest falcons are a small group of birds that live in the forests of South America. The first study on the breeding biology of this group was only undertaken in 1992, so it is hardly surprising that little is known on these birds. They all look somewhat 'hawk like,' they all nest in tree cavities and lay whitish eggs rather than the typical brown mottled eggs of the falcons and caracaras. They feed on a wide variety of forest fauna - insects, birds, snakes, reptiles and possibly small mammals, although with several species virtually nothing is known.

The smallest of all the diurnal raptors are the falconets. One species comes from South America - the Spot-winged Falconet, one species from Africa, originally known as the African Pygmy Falcon and six from south east Asia. They all nest in holes or old bird nests, they lay white eggs. Most are quite vocal and perform a visual display with head bobbing and tail flipping. They feed mainly on large insects, small birds, reptiles and the occasional mammal. The smallest weighs under 50 grams and the young hatch at under 5 grams.

The last group in the diurnal raptors is the true falcons. These contain the well-known Peregrine and the Kestrel, the ladies falcon - the Merlin and many others. There are 13 species of Kestrels, most of these little birds live in a wide variety of habitats including towns, farm land woodland edges and the more open areas. Most live on insects, reptiles and small mammals and most can and do hover.

Peregrine Falcon *(Falco peregrinus)*

The other group of smaller falcons include the merlins, hobby's, new zealand falcon and the all australian falcons except for the peregrine. These birds do not have the ability to hover like the kestrels and most live in the more open areas of the world.

The large falcons are all birds of the open plains, mountains, deserts, savanna and similar terrain. Being larger they hunt larger prey and most live on birds or mammals with occasional forays on insects when termites start to fly. They have weak feet in comparison to the hawks, but the beak is very powerful. If quarry is not killed instantly by the speed of the stoop, the falcons have a special tooth on the beak called the tomial tooth, this is used to break the neck of the quarry. Nearly all the falcons have long pointed wings and fairly short tails and most have dark brown eyes. None of the falcons build their own nest, but either dig a scrape in the ground or on cliff ledges or even in the build up of dirt on high rise building ledges. Some choose holes in trees and some will use bird nests.

Lanner Falco (Falco biarmicus)

Common Kestrel (Falco tinnunculus)

Merlin (Falco columbarius)

Nocturnal Birds of Prey (Owls)
(Scientific group name **Strigiformes**)

The Barn Owls
(Scientific family name **Tytonidae**)

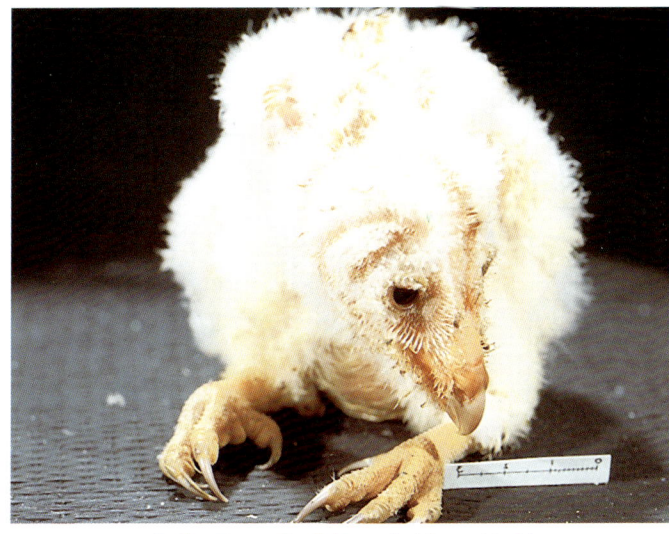

Probably the best known and loved of all the owls are the barn owls. They are found throughout the world, with different species living in different areas and habitats, but all looking very like the Barn Owl we know. Of all the owls, they have the most marked facial disc giving them the heart-shaped look that they all have. Several of the Barn Owl species are rare - mainly because their habitat is reducing and this is affecting them badly. Like all the owls they use their hearing as much if not more than their eyesight to locate their prey. It is said that a barn owl can locate a mouse in zero light - i.e. in pitch darkness using only its hearing to locate and catch the moving mouse.

Between the barn owls and all the other owls are two species of owls called Bay owls. One comes from the far-east and one from Africa - there is very little known on the African bird, apart from one dead specimen and more recently some owls heard vocalising. The Oriental Bay Owl is better known. Smaller than the barn owls and the facial disc is more square than heart-shaped, these are forest owls a little bigger than the European little owl.

Baby Barn Owl *(Tyto alba)* 3 weeks old

Barn Owl *(Tyto alba)*

The Rest of the Owls
— sometimes known as the Eared-Owls
(SCIENTIFIC FAMILY NAME **STRIGIDAE**)

All the other species of owls are put under one family name although they vary a great deal and are divided into two sub families.

The Buboninae contains small to tiny owls such as the White-faced Scops owl from African and the European Scops owl, which only weighs about 60 grams. These Scops owls form the biggest group with 34 species. Also in the same sub-family are the eagle owls, which are large to very large owls, all with feather ear tufts. The biggest of all is the Eurasian Eagle owl and the females can go up to 9lbs in weight.

The fishing owls look similar to the eagle owls although they tend to have longer ear tufts, but they don't have feathered legs and toes, as they drop their feet into the water when fishing. Thus their legs and feet are kept drier.

The tiniest of all the owls sit in this subfamily - the pygmy owls, the Least Pygmy Owl has a descriptively scientific name - *Glaucidium minutissimum*. All the hawk owls come under the same grouping but these are divided. There is one northern Hawk Owl - you can see it at the Centre, all the others come from the southern-hemisphere. As their name denotes they have a hawk-like build with the same long tail that typifies the hawks. The northern hawk owl flies in a very hawk-like manner and will happily hunt in the full daylight during the summer months. The Australian ninox owls are in this group too.

All the Australian owls are either Barn Owls (tytos) or belonging to the group (genus) ninox and they have some very odd ones, in two species the males are bigger than the females - which is unusual - mostly its the other way round. The female of the Sooty Owl on the other hand has a huge size difference between the female and male. In some races of this species, males are so tiny, they were even thought to be a different species.

Owl courtyard

Tawny Owl
(Strix aluco)

Great Grey Owl
(Strix nebulosa)

The other sub-family in the owls is the Striginae. Our own well-known and commonest owl - the Tawny Owl belongs here, along with the Barred Owl, Great Gray Owl and the beautiful Asian Wood Owl. Also found in the UK are the Short-eared and Long-eared owls, which come in this group. The Fearful Owl from the Solomon Islands is a relatively small owl, but with a very powerful beak and huge feet for its size, it is said to be as powerful as a Great Horned Owl.

The last of the owls in the sub-family are four tiny high forest owls, which includes the Boreal Owl, also known as Tengmalms owl inhabits the cool forests of europe and north america.

Short-eared Owl *(Asio flammeus)*

Migration

Many birds move around the world on a yearly basis, breeding in one area and wintering in another. Sometimes the move will be driven by weather conditions, usually extreme low temperatures during the winter. Sometimes the length of daylight hours will encourage a bird to migrate to another country so it has extra hours in the day to hunt and feed its young, sometimes it is the food supply itself.

Certain raptors, such as the Steppe Eagle migrate very impressive distances - a young steppe eagle on its first migration flies approximately 8,000 miles south to southern Africa, to winter there, returning the same distance in the spring, covering a staggering 16,000 miles on the journeys. Some Ospreys migrate to the UK to breed as do Hobby's and Honey Buzzards, after the breeding season they and their young migrate south either to southern europe or Africa for the winter feeding grounds.

The routes followed are fairly well established for most land birds, although there is new research in progress using satellite radio tracking to follow individual radio tagged birds and see where they really go, what routes they travel and which populations use which areas for wintering. Birds of prey tend to follow land rather than crossing large areas of open sea. So, for example to cross the Mediterranean they follow routes that have what is called 'land bridges' crossing at the narrowest point, not necessarily the straightest route.

This does lead to problems because it funnels large numbers of birds through fairly tight corridors. Sadly less pleasant human populations around the world take advantage of this and in some areas they shoot high number of birds of prey and other birds on migration. One of the worst culprits is Malta which has the dubious honour of shooting as many as five million birds a year, including over a million swallows and 60,000 birds of prey and 20,000 owls. They even go out in boats to shoot the birds coming in over the sea.

Saker Falcon
(Falco cherrug)

Slowly people are beginning to understand that this is a vile and useless practice, but progress is slow. Now in some areas of Italy the problem is reducing, education is the key to try and teach the young people that this is a pointless thing to do and can badly affect wild populations of many species of birds.

So spare a thought when you see Martins, Swallows and Swifts arrive each spring or you are lucky enough to have Hobbys around where you live. Imagine the huge distances, the weather conditions, the dangers faced, and the energy and drive they have, to achieve this incredible feat each year in order to breed.

Ferruginous Buzzard
(Buteo regalis)

Identification

Time and time again here at the Centre we have people phone us to ask if we have lost an eagle. Now occasionally we do lose birds here and we are delighted when people spot them and let us know, but more often than not the bird will be a wild one. There are a number of things to remember when spotting wild birds of prey. Firstly they behave differently at different times of the year. For example it is a common sight to see Buzzards wandering around on the ground in fields in the winter and early spring. They may even be in the same field for several days. At that time of year they eat a lot of grubs, beetles and other insects.

They are not waiting for lambs, although they may eat scraps of after birth. In fact they often do farmers a great deal of good by eating pests that may damage crops later in the year.

In cold weather some birds of prey are likely to come into your gardens - particularly if you have a bird table - a bird table for small birds is a bird table for sparrowhawks and even kestrels. Tame white pigeons are also very tempting to wild birds of prey, especially when the weather conditions are hard for them.

So where you might not see birds of prey close to in the summer, you are more likely to do so in the winter. The other thing to remember is that not all birds of the same species look the same. They have adult and juvenile plumage, sometime sub adult as well, all of which is different. In some species the male is a different colour from the female, and a different size. And in other species there is a great deal of variation in colouration from one individual to another. So you may look in your bird book and see a photo or painting of a buzzard and the bird that you are looking at is not remotely the same colour. Buzzards can and do vary from dark dark brown, through all sorts of different mottled browns, to occasionally almost white.

What should inform you which family the bird belongs to, is the silhouette. Remember - usually females are bigger than males. Falcons in the UK all have brown eyes, hawks yellow or orange. Four out of the six owls have yellow eyes. Check that Snowy Owls for example come from the area you are seeing the bird, they are an arctic species so we are unlikely to have them in Gloucestershire. The amount of times we have been called out to check on a definite sighting of an eagle, is countless - never has it been an eagle - its always been a buzzard. Common Buzzards have a wingspan of four and a half feet and they look very big, especially when seen close and low. Golden Eagles have a wing span of eight feet - you will never forget it if you genuinely see one of these close up.

A note here – recently birds of prey have been accused of causing the decline in small garden bird populations in the UK. This is in fact demonstrably not true. The changes in the face of the countryside where the small birds used to live, removal of hedgerows, ploughing of stubble fields in the summer instead of leaving them over the winter and ploughing in the spring, thus removing a huge food resource. Over grazing, pesticide sprays removing the food source, all have helped the decline. Even bird tables that are piled with food but rarely cleaned, and so can and do spread disease. All of these are the true reasons for the decline, not birds of prey.

Some of the other work at NBPC

Apart from what you will see on a visit here to NBPC, there are many things that you don't see, for example:

Work Experience and Training

Some of the people seen working here may well be either volunteers or work experience. We take on many work experience students over the year. Apart from the usual school students who are only here for a week, we also have students from Animal Management Courses and Veterinary Colleges. We are delighted to have these students here from four to twelve weeks, it is satisfying to know that young vets will start out at least recognising a bird of prey, understanding one end from the other, and being able to handle birds with confidence. The same can be said of students who are going into animal management. They can only benefit by understanding birds of prey a little better and that is good for them and the birds.

At the same time, in training, we run five day Falconry Courses and three day Owl Courses. Of particular interest to us, are the Police Wildlife Liaison Officer Courses, which allow us to help police officers to have a greater understanding of birds of prey, better powers of identification and handling techniques. We run similar courses for RSCPA Wildlife staff and those dealing with Imports.

Rehabilitation

Every year we have between 50 and 100 injured wild birds of prey brought to the Centre. We advise on probably double that number over the phone. These birds are treated here, or with our vet and where possible they are returned to the wild where they belong.

Assistance abroad

For some years now the work at the Centre has expanded to internationally, we travel to various places and using the expertise learned over the three decades here, assist where we can.
Egyptian Vulture Breeding Project – Italy
Indian White-backed Vulture Project – Bombay and National Parks India
Philippine Eagle Project – Mindanao
Spanish Imperial Eagle Project – Spain
South Carolina Center for Birds of Prey – US

Not only do we visit the projects and sites, but also try to provide training in management techniques on site at The National Birds of Prey Centre

Advice

We have upwards of 80 telephone calls a day, many of which are people wanting advice which we freely give, although it is not always the advice they might want. We also have about 30 – 40 daily emails and letters, many of these are from people wanting help. We prefer the emails as they are easier to deal with. However, please check our web-site first – you may find the answers there. If you are writing to us it is helpful if you put in an stamped addressed envelope. Don't send Faxes, the director hates them and answers them last of all, if at all.

The Breeding Successes

Falcon eggs

We started breeding birds in the early 70's, although Kestrels bred here in 69 two years after we opened. Then followed Lanners and Ferruginous Hawks, Red shouldered Hawks and Caracaras. Once we started to really understand what we were doing successes followed thick and fast until now today we lead the world in the captive breeding of different species. Up to the year 2000 we have bred 60 species of diurnal and nocturnal birds of prey.

The Breeding Barns

When we started to replace all the old pens with Barns, we did try to limit each barn to a specific family, for example, eagles in the Eagle Barn, falcons in Barn 3 and so on. However it did not quite work out like that. Some of the birds did not like being to close to the flying ground, as being territorial in the breeding season, it disturbed them to see the free flying demonstration birds near their 'territories', and they failed to breed. A few of the birds did not like to be able to see others of the same species. Some birds like the quieter parts of the Centre, others like the female Andean Condor, likes to be where there are most people. Some believe it or not won't breed in the larger pens but will in a smaller one. So slowly over the years we have moved things around so there is no particular order. This can be annoying for us, and the visitor, however as one of our prime reasons for being here is to continue breeding birds, we try to fit in with their needs.

One major problem is that as soon as one becomes successful at breeding there is instantly far more space needed. It is only sense to start to put together young birds to replace your older breeding birds once they get too old. We always need more space to build more pens and we have some new designs for pens that we would really like to try, and will once we have the finances to do this.

African Pygmy Falcons *(Polyhierax semitorquatus)*
and
Indian Egyptian Vulture *(Neophron percnopterus ginginianus)*

The resulting young

One often answered question is 'what do we do with the birds that we breed' Many ask if we release them to the wild. Generally the answer is no. For a release programme to be successful, firstly the birds must have been there historically, secondly the reason they are no longer there has to have been solved. Then for the released birds to survive in enough numbers to breed themselves, and with enough genetic variety, we would have to be breeding from a minimum of four pairs of the species we are wanting to release (depending on the species), and ideally many more. This is not what we are able to do here. Although we are working towards breeding and releasing Red Kites in the next few years. Our brief is sustained captive breeding over many years, and producing what are called F2 birds. These are young bred from birds already bred in captivity. Once that has been achieved, one really does begin to understand the process and become able to breed birds more easily and learn more.

Although the young are rarely released to the wild, they do have a purpose. We use a number of birds here on demonstration yearly, we are always adding to the team as the older members reach retiring age, or want to breed. We also like to add in new species that we have not flown before and we are keen that these are wherever possible, birds that we have bred ourselves. Some of our young go to other collections and keepers, some go abroad, some may go to breeding programmes where their young will be future releases into the wild. Anyone who has a bird from us goes through rigorous questioning and needs to be able to prove in writing and photographs, that they are able to look after and fly the bird and the housing is more than acceptable. And Yes, we have turned people down.

Species of raptors bred at NBPC 1967 - 2000

Turkey vulture	Cathartes aura
American Black Vulture	Coragyps atratus
Red Kite	Milvus milvus
Yellow-billed Kite	Milvus migrans parasitus
African Fish Eagle	Haliaeetus vocifer
Bald Eagle	Haliaeetus leucocephalus
Egyptian Vulture	Neophron percnopterus
African Harrier Hawk	Polyboriodes typus
Gabar Goshawk	Micronisus gabar
Northern Goshawk	Accipiter gentilis
European Sparrowhawk	Accipiter nisus
Black Sparrowhawk	Accipiter melanoleucus
Harris Hawk	Parabuteo unicinctus
Common Buzzard	Buteo buteo
Red-tailed Buzzard	Buteo jamaicensis
Ferruginous Buzzard	Buteo regalis
Red-shouldered Buzzard	Buteo lineatus
Tawny Eagle	Aquila rapax
Golden Eagle	Aquila chysaetos
Verreaux Eagle	Aquila verreauxii
Blyth's Hawk Eagle	Spizaetus alboniger
Secretary Bird	Sagittarius serpentarius
Common Caracara	Polyborus plancus
Striated Caracara	Phalcoboenus australis
African Pygmy Falcon	Polihierax semitorquatus
Kestrel	Falco tinnuculus
American Kestrel	Falco sparverius
Mauritius Kestrel	Falco punctatus
Merlin	Falco columbarius
Aplomado Falcon	Falco femoralis
Peregrine Falcon	Falco peregrinus minor
Saker	Falco cherrug
Prairie Falcon	Falco mexicanus
Lugger Falcon	Falco jugger
Lanner Falcon	Falco biarmicus
Hybrid Falcons	Falco hybridi
Barn Owl	Tyto alba
Tawny Owl	Strix aluco
Great Gray Owl	Strix nebulosa
Asian Brown Wood Owl	Strix leptogrammica
Rufous-thighed Owl	Strix rufipes
Little Owl	Athene noctua
Snowy Owl	Nyctea scandiaca
Spectacled Owl	Pulsatrix perspicillata
European Eagle Owl	Bubo bubo
Bengal Eagle Owl	Bubo bengalensis
Great Horned Owl	Bubo virginianus
MacKinder's Eagle Owl	Bubo capensis
African Spotted Eagle Owl	Bubo africanus
Savigny's Eagle Owl	Bubo ascalaphus
Abyssinian Eagle Owl	Bubo cinerascens
Iranian Eagle Owl	Bubo b. nikolskii
White-faced Scops Owl	Otus leucotis
European Scops Owl	Otus scops
Tropical Screech Owl	Otus choliba
Collared Scops Owl	Otus lempiji
Boobook Owl	Ninox novaeseelandiae
Burrowing Owl	Athene cunicularia
Striped Owl	Rhinoptynx clamator
Hawk Owl	Surnia ulula

Some stories about special birds here

At least half the birds here have names, this is because either they have been special to us in some way, or they were, or still are, a part of the flying team. Some came to us in odd circumstances, others have been ill, had a lot of time and effort spent on them and recovered. I should add here that if a bird does not have a name - it is still just as important to us.

Some of the birds have been here a long time and are pretty old. Some like one of the African Fish Eagles are particularly difficult, especially during the breeding season.

One of the oldest pairs of birds here are Mr and Mrs Timms - called after the person who owned them before us. They are a pair of Indian Tawny Eagles and are beginning to show their age now. We think that they are around 47 years old, possibly older. They have bred at least 20 young, four of which are still with us here. They taught us how to deal with young eagles, both in rearing and training, they made history by rearing three young one year (usually this species of eagles only rears one young because the oldest baby kills the youngest – a habit called Cainism). Mrs Timms nearly died on us in 1996, the previous year she bred one male that we kept here - you may see him fly in the summer, his name is Hard Tackle. We thought that she was unlikely to breed again, not really surprisingly, but to our amazement on January 9th 1998 she laid an egg again. At the time of writing this new guide, January 2001 she is sitting on two eggs which we hope will be fertile and will hatch. There is no doubt she and her mate are pretty remarkable. But whatever the outcome we still love her and they seem to still be enjoying each other and lead a happy life. The normal life span for this type of bird would probably be around 15 - 20 years. We feel very strongly here, that even though a pair are no longer productive, they still deserve a good retirement after all their efforts for us.

Mozart, is another aging bird. He is the European Eagle Owl who lives at the end of the Hawk Walk. He was born in 1973. He spent his childhood at the Royal Academy of Music until he could fly. Then he came home. In 1981 he broke his jesses and went missing in one of the coldest winter spells since 1949, he was out for 71 days. Then in the early spring, all by himself, he came home again. How he survived we have no idea. He is one of the most popular birds here, particularly with the parties of visually impaired children as he will let them stroke him. Each spring he tries to get the staff to breed with him and in the evenings he will offer us a dead rat through the wire to prove what a good husband he would be. His mother lived until she was about 25. At the time of writing his father Robert is still alive at 35.

Sole arrived here in 1996. She is an albino Kestrel, we had never seen one in all the years of the Centre. Ironically in the same year she came in an albino Tawny Owl was brought in. The owl is not a white as Sole - more a sandy brown, but he has the pink eyes of an albino. Neither bird could survive in the wild. Luckily for Sole she was found before she died of starvation, but even now if we fly her outside the field she gets attacked by other birds. She earns her keep by flying on demonstration and rests in a pen in her resting, moulting period. Sadly she is a bad tempered little bird and can't be kept with other birds as she is very aggressive to them. But on a sunny day to see her fly on demonstration with the sunlight gleaming through her white wings is a joyful sight.

Brown Argus is a female Red Kite, she was rescued from a nest in Wales and taken to a Centre to be reared as she was extremely ill. Once grown she was sent to the Northampton release programme in 1998. There she was released – several times, and she did not do well, in the end it was decided to send her here for us to assess. We trained and flew her, but she behaves in a manner that lead us to know she got too accustomed to humans in the early stages of her life and when under pressure, such as getting hungry she resorts to asking humans for food. This is not good practice in a wild bird! Leads to all sorts of problems, so she is here. She will eventually be given a mate and in the future her offspring will be released back to the wild.

Questions for you to answer

Using this guide, the information in the education room and around the Centre, plus what you hear on the demonstrations try answering the following questions just for fun

1. What are some of the differences between the Old World and the New World vultures?
2. How can you tell a falcon from a hawk - name three differences?
3. What are the two main features that make a bird of prey different from other meat eating birds?
4. Which is one of the rarest vultures in the world?
5. Name two types of birds of prey that eat snakes
6. Which two family groups of birds of prey do not build a nest?
7. Name one of the species of raptors that lays over seven eggs?
8. Which bird of prey lives on every continent except Antarctica?
9. Choose two species of birds and name the habitat that they live in
10. What is the leather cap some of the birds wear on the demonstrations?

(answers on page 31)

Little Owl
(Atheni noctua)

How You Can Help

By visiting us here at The National Birds of Prey Centre and buying this guide, you have already assisted us in a small way to continuing our work. We hope that by the time you have seen around The Centre and watched some of the birds fly, you will understand why we are here and what our aims are. If, at the same time as having had an enjoyable day, you have learnt a little about birds of prey, then we have done our job. It is vital that we all have a greater understanding and affection for the other species which inhabit our planet, if we are to conserve what remains.

I would like to thank you for visiting. Please continue to support us by coming again. There are many other ways you can help us.

- Take a leaflet and give it to a friend who might enjoy a visit
- Take a leaflet and join as a member of the centre
- Take a leaflet and sponsor a bird
- Come on one of the Falconry Experience Days or courses
- Donate plants, tools, or equipment.
- Donate time or professional advice
- Donate money
- Leave our Charity, The National Birds of Prey Trust, or the Centre itself a bequest in your will

Should you wish to help us in any other way, please do not hesitate to ask one of the staff, or write to me here at the Centre. Conservation is a costly and precious business and we need all the help we can get.

Jemima Parry-Jones MBE
Director

Pleasant grounds in which to stroll

Courses held at the National Birds of Prey Centre

FALCONRY EXPERIENCE DAY

Gives you the chance of close personal contact with various species of Raptors, including many British species. Starting at 9.30am you will learn t fly birds to the fist, and watch a special demonstration. You will then be given expert tuition in handling some of our many trained birds, including Falcons and an Eagle.

After lunch here at the Centre, you will have a little time to look around, whilst the working birds are loaded into the Land Rover. The day culminates in an afternoon our hawking in the local countryside with course members able to see trained birds exercised, and flown at quarry. Rest periods sitting under a tree and watching the wildlife, including wild birds of prey, makes a memorable afternoon.

On you return there may be time to watch the last flying demonstration at the Centre or have a browse in the Gift Shop before going home. This is not a full Falconry Course, but is designed to be an insight into birds of prey and the world of falconry. Treat yourself to a special day, or give as a unique gift to friends and family. Falconry Experience Days are held throughout the year, except January. Vouchers (valid for one year) are available by mail order, and the date can be organised by the recipient.

FIVE DAY FALCONRY COURSES

Are designed for the those who want to learn enough to own and train a bird for falconry. It covers; * The different and most suitable species for falconry * Initial training, * Flying and hunting * housing * management * veterinary care * biology * equipment * telemetry * lure swinging, and other falconry skills. There are at least two afternoons out hawking, sometimes more depending on the weather. It is a very intensive week with the emphasis being hands-on experience. Classroom instruction is given as well as field craft skills. This is **the** course for people who are really serious about owning their own bird. Five Day Falconry Courses are held September to May. This course is very intensive and requires hard work on your part. When booking you will be sent information on clothing needed, useful books, videos and a timetable, which will help you get more from the course. A deposit is required when booking.

THREE DAY OWL COURSES

As far as we know this is the only course run on the management and training of captive owls. We have designed this course to give you an insight in to the keeping of an owl safely and well. There is a mistaken idea that training owls is easier than training other birds of prey, they are not!

During the course you will be working with young owls at the Centre. This is why we prefer to run it during the breeding season, when we have young birds. You will learn how to handle, manage, feed, train and fly an owl. You may decide at the end of the course that an owl is not for you, that is a good thing, as it is wrong to get a bird and then find out it is not for you.

CORPORATE DAYS, SPECIAL DAYS AND EVENINGS EVENTS

The Centre can provide a wide range of events which can include activities outside falconry. If you require an unusual day or evening for business colleagues, club, family or friends, contact us and we will design a day or evening including refreshments to meet your needs and your pockets. Anything from 6 to 600 depending on what you want. We can work from here, from local known venues with top quality facilities, even come to a venue of your choice as long as the space is sufficient.

Mail Order Services

We run a very efficient mail order service from the National Birds of Prey Centre. In addition to Falconry Experience Day Vouchers, we also have a number of books, videos, gifts and clothing with a bird of prey or owl theme available. For up-to-date details of what is available, please check our website: www.nbpc.co.uk

It's easy to order, just telephone us on 0870 990 1992 (Fax: 01531 821389 or Email: jpj@nbpc.demon.co.uk) during office hours or leave a message on our answering service.

If you have any questions or would like more information on MEMBERSHIP, BIRD SPONSORSHIP, CHILDREN'S BIRTHDAY VISITS and other gift ideas please contact us.

Further Reading

Owls

Allan W. Eckert. **THE OWLS OF NORTH AMERICA.** Doubleday and Co. New York

Heimo Mikkola. **OWLS OF EUROPE.** T & AD Poyser. Calton.

D.S. Bunn. A. B. Warburton. R.D.S. Wilson. **THE BARN OWL.** T & AD Poyser. Calton

Varying Authors. **THE BIRDS OF NORTH AMERICA.** (The varying American owls, published in individual species papers). The Academy of Natural Sciences of Philadelphia.

John Burton. **OWLS OF THE WORLD.** Peter Lowe, Eurobook Limited.

Krystyna Weinstein. **THE OWL IN ART MYTH AND LEGEND.** Grange Books Published by Universal Books London.

Paul A. Johnsgard. **NORTH AMERICAN OWLS.** Smithsonian Institution Press. Washington and London.

John Sparks and Tony Soper. **OWLS.** David and Charles. Newton Abbot.

R Hume. **OWLS OF THE WORLD.** Dragonsworld Book. Limpsfield. London.

Parry-Jones J (1998) **UNDERSTANDING OWLS.** David and Charles.

Diurnal Birds of Prey

Freethy, Ron. (1982) **HOW BIRDS WORK.** A guide to Bird Biology. Blandford Press. Dorset. UK.

Brown, L. & Amadon, Dean. (1968) **EAGLES, HAWKS AND FALCONS OF THE WORLD.** Country Life Books, distributed by Hamblyn Publishers.

del Hoyo, Josep. & Elliott, Andrew. & Sargatal, Jordi - Editors. (1994) **HANDBOOK OF THE BIRDS OF THE WORLD. VOLUME 2 NEW WORLD VULTURES TO GUINEAFOWL.** BirdLife International published by Lynx Edicions.

Parry-Jones J (1987) **FALCONRY, CARE CAPTIVE BREEDING AND CONSERVATION.** David and Charles.

Parry-Jones J (1992) **TRAINING BIRDS OF PREY.** David and Charles.

Quiz Answers

1. The old world Vultures are related to Eagles the New World Vultures are related to Storks.
2. Falcons have long pointed wings, short tails and brown eyes, Hawks have short rounded wings, long tails and yellow or orange eyes.
3. They catch their food with their feet, rather than their beaks as other birds do.
4. The rarest Vulture is the Californian Condor (at the time of writing).
5. Snake Eagles and Secretary birds, Some buzzards will and so do the White-Bellied Sea-eagles.
6. The Falcons and Owls.
7. Snowy Owls, Short-eared Owls, Rough-legged Buzzards.
8. Peregrines and Ospreys.
9. That's up to you to choose and get right!
10. A hood to keep them calm.

The National Birds of Prey Centre is only as good as it is, because of all my staff . . .
Thank you — JPJ

Useful contacts and addresses

LICENSING AND REGISTRATION AUTHORITIES

Department of the Environment, Transport and the Regions (DETR)
Wildlife Licensing Department
Tollgate House
Houlton Street
Bristol BS2 9DJ
Tel: 0117 987 8686

IMPORT/EXPORT HEALTH CERTIFICATES ETC

**Ministry of Agriculture, Fisheries & Food (MAFF)
Import / Export Section**
Area 402 1A Page Street
London SW1P 4PQ
Tel: 020 790 46352/46353

VETERINARY SURGEONS SPECIALISING IN RAPTORS

N. Forbes, MRCVS, B.Vet.Med.
Lansdown Veterinary Group
The Clockhouse Veterinary Hospital
Wallbridge
Stroud
Gloucestershire GL5 3JD
Tel: 01453 752555

A. Greenwood
International Zoo Vet Group
Keighley Business Centre
South Street
Keighley
West Yorkshire BD27 1AG
Tel. 01536 560000

N. Harcourt Brown BVSc
30 Crab Lane
Bilton
Harrogate
North Yorkshire HG1 3BE
Tel: 01423 508945

S. Spencer BVSc MRCVS
Maison Dieu Veterinary Centre
76-77 Maison Dieu Road
Dover
Kent CT16 1RE
Tel: 01304 201617

J. Chitty
Strathmore Veterinary Clinic
No 6 London Road
Andover
Hants SP10 2PH
Tel: 01264 352353

CLUBS AND ASSOCIATIONS

The Hawk & Owl Trust
B. Winder
Information Officer
51 Eton Wick Road
Windsor
Berkshire SL4 6LX
Tel: 01753 854393

The Hawk Board/Welsh Hawking Club
M. Clowes
10 Birthorp Road
Billingborough
Lincolnshire NG34 0QS
Tel: 01529 240443

British Falconers Club
J. R. Fairclough
Home Farm
Hints
Nr Tamworth
Staffordshire B78 3DW
Tel: 01543 481737

The Countryside Alliance (BFSS)
367 Kennington Road
London SE11 4PT
Tel: 020 7582 5432

OTHER FALCONRY CENTRES

The Hawk Conservancy
Weyhill
Nr Andover
Hampshire SP11 8DY
Tel: 01264 772252

FALCONRY COURSES

The National Birds of Prey Centre
Newent
Gloucestershire GL18 1JJ
Tel: 0870 990 1992

FALCONRY EQUIPMENT

Falconry Furniture
Parsonage Farm Buildings
Llanrothal
Monmouth
Gwent NP5 3QJ
Tel: 01600 750300

- Eagle
- Hawk (Accipiter)
- Buzzard (Bueto)
- Falcon
- Harrier

The National Birds of Prey Trust

Promotes the conservation, protection and preservation of all species of raptors, especially, but not exclusively, through

> **Education**
> **Training**
> **Captive Breeding &**
>   **Re-introduction programmes**
> **Rehabilitation programmes**
> **Research**

The Trust achieves it's objectives through:

EDUCATION

The central and crucial objectives, without an understanding of why raptors are so important, the other aims will fail

> PUBLICATIONS
> Educating a broad base with the general public, particularly the education of school children and, even more importantly, their teachers. The Trust will produce a Schools pack on birds of prey, tailored to fit in with the National Curriculum, which will be distributed free of charge to schools.

NATIONAL AND INTERNATIONAL TRAINING

The training of national and overseas students is already in progress. Students from all over the world need to come and study at internationally respected, leading British establishments. However, travel and housing is expensive. Students do not have the funds to support themselves, this is particularly relevant for those from the developing world, which often means they cannot receive training. The Trust is their hope. For example in India, raptors, particularly the vultures, vital to the eco-system are fast disappearing. The Philippine Eagle Foundation at Daveo is dependant on its public education and needs the secondment of staff from Britain to assist with their work. Staff require help and training to ensure that the wild stock is maintained through habitat management Education programmes need to be put in place, and when necessary, breeding programmes initiated or assisted. All this work has to be funded almost entirely by voluntary subscription.

CAPTIVE BREEDING & RE-INTRODUCTION PROGRAMMES

In the United Kingdom the translocation and release of young RED KITES has been a spectacular success. The scheme can now expand to other areas. A captive breeding programme has been set up to produce young birds, starting with a release scheme in Gloucestershire, in partnership with English Nature.

Internationally the calls on the Trust are world wide.

The Spanish Imperial Eagle is one of the most endangered species of eagle left in the wild. In conjunction with the Spanish Government and scientists working on the project, the Trust has been asked to help with expertise and therefore funds to preserve the species from extinction.

Egyptian Vultures are another endangered species within Europe. With WWF of Italy, a captive breeding and release programme is underway following training undertaken in the UK.

There has been a 90% reduction in the population of the Indian White-backed and Long-billed Vultures over the last 10 years, almost to the point of no return. Another call for help received recently by the Trust.

REHABILITATION OF INJURED WILD BIRDS

Covering the high costs of housing, treating, housing and managing injured wild raptors is expensive. Helping to meet those costs should be a vital part of our work. Currently nearly 400 birds are dealt with yearly at one centre alone. The calls for help are far great but cannot be met due to lack of funds.

RESEARCH

The Trust employs no staff of its own, it calls on the facilities and expertise of others, all of which needs to be paid for. Many calls for much needed research are received. Few can be answered without greater funding.

DNA studies - *ongoing*

Studies into bird immune systems - *completed*

Food aversion therapy studies with falcons - *ongoing*

Studying flight performance of migratory species - *completed*

Studies on the gut performance of sedentary and active raptors - *completed*

Radio tracking of Swallow-tailed Kites on migration - *ongoing*

Many of the Trust's projects require no more than £1500 to set them up. Other larger and more ambitious vital projects need a great deal more. Research and rehabilitation takes time and usually spans many years.

The Trust through the support of its Friends can and will be able to help save many endangered species. Your contribution is vital, however small or large, as a lump sum or as yearly contributions. Please help as much as you can, for example

£500 will pay the travel for an overseas student or a month of living costs

£1,000 covers the cost of an injury rehabilitation for 6 months

£5,000 will help fund a research programme

£10,000 will set up an important breeding programme

The National Birds of Prey Trust employs no staff, it is entirely dependant on voluntary support and contributions.

Glossary

A.I. - Artificial Insemination - the fertilising of a female bird prior to egg laying by placing semen into her at the correct time.

Booze - When a bird of prey takes a drink – that's where the term 'the boozer –i.e. the pub comes from!

Bow Perch - The type of branch like perch that hawks and buzzards prefer.

Block - The type of perch that a Falcon prefers.

Breeding Season. - The time of year when birds breed - usually the spring in Britain, although many of our foreign birds breed at very odd times of the year!

Bewitt - The strap used to hold the bell on the birds' leg.

Bath - The metal bowl used for the birds to bath in.

Brood - To sit on and keep eggs and the young birds warm.

Buzzard - A medium sized raptor with broad wings, generally living in more wooded country.

Cadge - A wooden frame perch which can be carried with shoulder straps for travelling trained birds – this is probably where the name 'golf caddy' came from.

Crepuscular - A bird which flies at dawn and dusk in the half light.

Carrion - An animal or bird that is dead.

Carcass - Body of dead animal or bird.

Captive Breeding - Breeding animals or birds in captivity.

Clutch - The size and number of eggs that a bird lays.

Close Ring - A seamless metal band or ring put on a bird's leg at about ten days old for identification.

Creanse - A long nylon line used to control a bird during the training period.

DETR - Department of the Environment, Transport and the Regions. The governmental body that legislates and controls the registration of certain birds of prey.

Diurnal - Coming out or living in the daylight.

Dummy Bunny - a rabbit shaped lure used to teach the bird to chase and hunt rabbits.

Eagle - Generally large birds living in more open countryside, with long and broad wings a longish tail and very powerful feet.

Eagle Owl - Large owls that have mistakenly been thought of as related to eagles - which they are not.

Falcon - Small to medium raptors, generally living in open countryside, with long pointed wings, shortish tails, brown eyes, and a 'Tooth' on either side of the hook on the beak.

Hybrid - a cross between two different species of birds.

Hawk - Small to medium birds with short rounded wings, long tails, usually yellow eyes, living in wooded countryside.

Hawk Walk - The area at the National Birds of Prey Centre, where all the trained birds are kept.

Hood - The leather cap or hat used, mainly on Falcons, to 'hoodwink' them into thinking it is night-time and therefore be calm.

Immature - A young bird in its first year's plumage.

Incubator - An electrical machine used to keep eggs warm enough to hatch without the mother bird.

Imprint - A bird that has been hand reared and is confused as to whether it is a bird or a human.

Juvenile - A young bird in immature plumage.

Jesses - The leather straps placed on the bird's legs.

Leash - The braided terylene line used to tether the birds.

Lure - The bird shaped item made of a pair of birds wings such as Moorhens and tied with a piece of meat and swung for a falcon to chase.

Mews - Indoor quarters for keeping birds of prey (from the French word muer - to moult).

Moult - The process of the bird losing and re-growing it feathers.

Nocturnal - A bird or animal that comes out after dark.

Owl - A bird, not related to other raptors but to Nightjars that hunts birds and animals, usually at night.

Pair Bond - The bird equivalent of marriage.

Recycle - When a bird either loses or has the first clutch of eggs removed and lays a second clutch.

Raptor - The more scientific name for a bird of prey.

Soar - To glide at a good height using hot air currents or winds.

Stoop - To drop or dive with folded wings from a height towards the ground.

Siblings - Brothers and/or sisters.

Swivel - The metal item that goes between the leash and the jesses to prevent any tangling.

Talons - Claws on bird of prey.

Tiercel - A male Peregrine (a 'tierce' or a 'third' smaller than a female).

Waiting On - When a bird circles above quarry waiting for it to flush.

Weathering Ground - An open area where a trained bird is put to get sunlight.